This planner belongs to :

_____

2026

January

S	M	T	W	T	F	S
				1	2	3
4	5	6	7	8	9	10
11	12	13	14	15	16	17
18	19	20	21	22	23	24
25	26	27	28	29	30	31

February

S	M	T	W	T	F	S
1	2	3	4	5	6	7
8	9	10	11	12	13	14
15	16	17	18	19	20	21
22	23	24	25	26	27	28

March

S	M	T	W	T	F	S
1	2	3	4	5	6	7
8	9	10	11	12	13	14
15	16	17	18	19	20	21
22	23	24	25	26	27	28
29	30	31				

April

S	M	T	W	T	F	S
			1	2	3	4
5	6	7	8	9	10	11
12	13	14	15	16	17	18
19	20	21	22	23	24	25
26	27	28	29	30		

May

S	M	T	W	T	F	S
					1	2
3	4	5	6	7	8	9
10	11	12	13	14	15	16
17	18	19	20	21	22	23
24	25	26	27	28	29	30
31						

June

S	M	T	W	T	F	S
	1	2	3	4	5	6
7	8	9	10	11	12	13
14	15	16	17	18	19	20
21	22	23	24	25	26	27
28	29	30				

July

S	M	T	W	T	F	S
			1	2	3	4
5	6	7	8	9	10	11
12	13	14	15	16	17	18
19	20	21	22	23	24	25
26	27	28	29	30	31	

August

S	M	T	W	T	F	S
						1
2	3	4	5	6	7	8
9	10	11	12	13	14	15
16	17	18	19	20	21	22
23	24	25	26	27	28	29
30	31					

September

S	M	T	W	T	F	S
		1	2	3	4	5
6	7	8	9	10	11	12
13	14	15	16	17	18	19
20	21	22	23	24	25	26
27	28	29	30			

October

S	M	T	W	T	F	S
				1	2	3
4	5	6	7	8	9	10
11	12	13	14	15	16	17
18	19	20	21	22	23	24
25	26	27	28	29	30	31

November

S	M	T	W	T	F	S
1	2	3	4	5	6	7
8	9	10	11	12	13	14
15	16	17	18	19	20	21
22	23	24	25	26	27	28
29	30					

December

S	M	T	W	T	F	S
		1	2	3	4	5
6	7	8	9	10	11	12
13	14	15	16	17	18	19
20	21	22	23	24	25	26
27	28	29	30	31		

Year in Pixels

	J	F	M	A	M	J	J	A	S	O	N	D
1.												
2.												
3.												
4.												
5.												
6.												
7.												
8.												
9.												
10.												
11.												
12.												
13.												
14.												
15.												
16.												
17.												
18.												
19.												
20.												
21.												
22.												
23.												
24.												
25.												
26.												
27.												
28.												
29.												
30.												
31.												

Color Codes

Notes

January

MONDAY	TUESDAY	WEDNESDAY	THURSDAY
			1
5	6	7	8
12	13	14	15
19	20	21	22
26	27	28	29

January

2026

FRIDAY	SATURDAY	SUNDAY	NOTES
2	3	4	○
			○
			○
			○
			○
9	10	11	○
			○
			○
			○
16	17	18	○
			○
			○
			○
			○
23	24	25	○
			○
			○
			○
			○
30	31		NOTES

February 2026

MONDAY	TUESDAY	WEDNESDAY	THURSDAY
2	3	4	5
9	10	11	12
16	17	18	19
23	24	25	26

February 2026

FRIDAY	SATURDAY	SUNDAY	NOTES
		1	○
			○
			○
			○
			○
6	7	8	○
			○
			○
			○
13	14	15	○
			○
			○
			○
			○
20	21	22	○
			○
			○
			○
			○
27	28		NOTES

March

MONDAY	TUESDAY	WEDNESDAY	THURSDAY
2	3	4	5
9	10	11	12
16	17	18	19
23	24	25	26

March

FRIDAY	SATURDAY	SUNDAY	NOTES
		1	○
			○
			○
			○
			○
6	7	8	○
			○
			○
			○
13	14	15	○
			○
			○
			○
			○
20	21	22	○
			○
			○
			○
			○
27	28	29	30
			31

April 2026

MONDAY	TUESDAY	WEDNESDAY	THURSDAY
		1	2
6	7	8	9
13	14	15	16
20	21	22	23
27	28	29	30

April 2026

FRIDAY	SATURDAY	SUNDAY	NOTES
3	4	5	○
			○
			○
			○
			○
10	11	12	○
			○
			○
			○
			○
17	18	19	○
			○
			○
			○
24	25	26	○
			○
			○
			○
			○
			NOTES

May

MONDAY	TUESDAY	WEDNESDAY	THURSDAY
4	5	6	7
11	12	13	14
18	19	20	21
25	26	27	28

May

FRIDAY	SATURDAY	SUNDAY	NOTES
1	2	3	○
			○
			○
			○
			○
8	9	10	○
			○
			○
			○
			○
15	16	17	○
			○
			○
			○
			○
22	23	24	○
			○
			○
			○
			○
29	30	31	Notes

June

MONDAY	TUESDAY	WEDNESDAY	THURSDAY
1	2	3	4
8	9	10	11
15	16	17	18
22	23	24	25
29	30		

June

FRIDAY	SATURDAY	SUNDAY	NOTES
5	6	7	○
			○
			○
			○
			○
12	13	14	○
			○
			○
			○
19	20	21	○
			○
			○
			○
			○
26	27	28	○
			○
			○
			○
			○
			Notes

July

MONDAY	TUESDAY	WEDNESDAY	THURSDAY
		1	2
6	7	8	9
13	14	15	16
20	21	22	23
27	28	29	30

July

2026

FRIDAY	SATURDAY	SUNDAY	NOTES
3	4	5	○
			○
			○
			○
			○
10	11	12	○
			○
			○
			○
			○
17	18	19	○
			○
			○
			○
24	25	26	○
			○
			○
			○
			○
31			NOTES

August

2026

MONDAY	TUESDAY	WEDNESDAY	THURSDAY
3	4	5	6
10	11	12	13
17	18	19	20
24	25	26	27

August

FRIDAY	SATURDAY	SUNDAY	NOTES
	1	2	○
			○
			○
			○
			○
7	8	9	○
			○
			○
			○
14	15	16	○
			○
			○
			○
			○
21	22	23	○
			○
			○
			○
			○
28	29	30	31

September

2026

MONDAY	TUESDAY	WEDNESDAY	THURSDAY
	1	2	3
7	8	9	10
14	15	16	17
21	22	23	24
28	29	30	

September

FRIDAY	SATURDAY	SUNDAY	NOTES
4	5	6	○
			○
			○
			○
			○
11	12	13	○
			○
			○
			○
18	19	20	○
			○
			○
			○
			○
25	26	27	○
			○
			○
			○
			○
			NOTES

October

MONDAY	TUESDAY	WEDNESDAY	THURSDAY
			1
5	6	7	8
12	13	14	15
19	20	21	22
26	27	28	29

October

FRIDAY	SATURDAY	SUNDAY	NOTES
2	3	4	○
			○
			○
			○
			○
9	10	11	○
			○
			○
			○
			○
16	17	18	○
			○
			○
			○
			○
23	24	25	○
			○
			○
			○
			○
30	31		NOTES

November 2026

MONDAY	TUESDAY	WEDNESDAY	THURSDAY
2	3	4	5
9	10	11	12
16	17	18	19
23	24	25	26

November 2026

FRIDAY	SATURDAY	SUNDAY	NOTES
		1	○
			○
			○
			○
			○
6	7	8	○
			○
			○
			○
13	14	15	○
			○
			○
			○
			○
20	21	22	○
			○
			○
			○
			○
27	28	29	30

December 2026

MONDAY	TUESDAY	WEDNESDAY	THURSDAY
	1	2	3
7	8	9	10
14	15	16	17
21	22	23	24
28	29	30	31

December

FRIDAY	SATURDAY	SUNDAY	NOTES
4	5	6	○
			○
			○
			○
			○
11	12	13	○
			○
			○
			○
			○
18	19	20	○
			○
			○
			○
			○
25	26	27	○
			○
			○
			○
			○
			NOTES

December
2025

01 MONDAY

02 TUESDAY

03 WEDNESDAY

04 FRIDAY

05 FRIDAY

○ _____
○ _____
○ _____
○ _____
○ _____
○ _____
○ _____
○ _____

06 SATURDAY

○ _____
○ _____
○ _____
○ _____
○ _____
○ _____
○ _____
○ _____

07 SUNDAY

○ _____
○ _____
○ _____
○ _____
○ _____
○ _____
○ _____
○ _____

08 MONDAY

○ _____
○ _____
○ _____
○ _____
○ _____

09 TUESDAY

10 WEDNESDAY

11 THURSDAY

12 FRIDAY

13 SATURDAY

○ _____
○ _____
○ _____
○ _____
○ _____
○ _____
○ _____
○ _____

14 SUNDAY

○ _____
○ _____
○ _____
○ _____
○ _____
○ _____
○ _____
○ _____

15 MONDAY

○ _____
○ _____
○ _____
○ _____
○ _____
○ _____
○ _____
○ _____

16 TUESDAY

○ _____
○ _____
○ _____
○ _____
○ _____

17 WEDNESDAY

18 THURSDAY

19 FRIDAY

20 SATURDAY

21 SUNDAY

○ _____
○ _____
○ _____
○ _____
○ _____
○ _____
○ _____
○ _____

22 MONDAY

○ _____
○ _____
○ _____
○ _____
○ _____
○ _____
○ _____
○ _____

23 TUESDAY

○ _____
○ _____
○ _____
○ _____
○ _____
○ _____
○ _____
○ _____

24 WEDNESDAY

○ _____
○ _____
○ _____
○ _____
○ _____

25 THURSDAY

26 FRIDAY

27 SATURDAY

28 SUNDAY

December
2025

29 MONDAY
- ○ _____
- ○ _____
- ○ _____
- ○ _____
- ○ _____
- ○ _____
- ○ _____
- ○ _____

30 TUESDAY
- ○ _____
- ○ _____
- ○ _____
- ○ _____
- ○ _____
- ○ _____
- ○ _____
- ○ _____

31 WEDNESDAY
- ○ _____
- ○ _____
- ○ _____
- ○ _____
- ○ _____
- ○ _____
- ○ _____

NOTES
_____
_____
_____
_____

01 THURSDAY

- ○
- ○
- ○
- ○
- ○
- ○
- ○
- ○

02 FRIDAY

- ○
- ○
- ○
- ○
- ○
- ○
- ○
- ○

03 SATURDAY

- ○
- ○
- ○
- ○
- ○
- ○
- ○
- ○

04 SUNDAY

- ○
- ○
- ○
- ○
- ○

05 MONDAY

○ _____
○ _____
○ _____
○ _____
○ _____
○ _____
○ _____
○ _____

06 TUESDAY

○ _____
○ _____
○ _____
○ _____
○ _____
○ _____
○ _____
○ _____

07 WEDNESDAY

○ _____
○ _____
○ _____
○ _____
○ _____
○ _____
○ _____
○ _____

08 THURSDAY

○ _____
○ _____
○ _____
○ _____
○ _____

09 FRIDAY

○
○
○
○
○
○
○
○

10 SATURDAY

○
○
○
○
○
○
○
○

11 SUNDAY

○
○
○
○
○
○
○
○

12 MONDAY

○
○
○
○
○

13 TUESDAY

14 WEDNESDAY

15 THURSDAY

16 FRIDAY

17 SATURDAY

○ _____
○ _____
○ _____
○ _____
○ _____
○ _____
○ _____
○ _____

18 SUNDAY

○ _____
○ _____
○ _____
○ _____
○ _____
○ _____
○ _____
○ _____

19 MONDAY

○ _____
○ _____
○ _____
○ _____
○ _____
○ _____
○ _____
○ _____

20 TUESDAY

○ _____
○ _____
○ _____
○ _____
○ _____

21 WEDNESDAY

- ○
- ○
- ○
- ○
- ○
- ○
- ○
- ○

22 THURSDAY

- ○
- ○
- ○
- ○
- ○
- ○
- ○
- ○

23 FRIDAY

- ○
- ○
- ○
- ○
- ○
- ○
- ○
- ○

24 SATURDAY

- ○
- ○
- ○
- ○
- ○

25 SUNDAY

○
○
○
○
○
○
○
○

26 MONDAY

○
○
○
○
○
○
○
○

27 TUESDAY

○
○
○
○
○
○
○
○

28 WEDNESDAY

○
○
○
○
○

29 THURSDAY

○ _____
○ _____
○ _____
○ _____
○ _____
○ _____
○ _____
○ _____

30 FRIDAY

○ _____
○ _____
○ _____
○ _____
○ _____
○ _____
○ _____
○ _____

31 SATURDAY

○ _____
○ _____
○ _____
○ _____
○ _____
○ _____
○ _____
○ _____

NOTES

_____
_____
_____
_____

February
2026

01 SUNDAY

○ _____
○ _____
○ _____
○ _____
○ _____
○ _____
○ _____
○ _____

02 MONDAY

○ _____
○ _____
○ _____
○ _____
○ _____
○ _____
○ _____
○ _____

03 TUESDAY

○ _____
○ _____
○ _____
○ _____
○ _____
○ _____
○ _____
○ _____

04 WEDNESDAY

○ _____
○ _____
○ _____
○ _____
○ _____

February 2026

05 THURSDAY

○ _____
○ _____
○ _____
○ _____
○ _____
○ _____
○ _____
○ _____

06 FRIDAY

○ _____
○ _____
○ _____
○ _____
○ _____
○ _____
○ _____
○ _____

07 SATURDAY

○ _____
○ _____
○ _____
○ _____
○ _____
○ _____
○ _____
○ _____

08 SUNDAY

○ _____
○ _____
○ _____
○ _____
○ _____

09 MONDAY

○ _____
○ _____
○ _____
○ _____
○ _____
○ _____
○ _____
○ _____

10 TUESDAY

○ _____
○ _____
○ _____
○ _____
○ _____
○ _____
○ _____
○ _____

11 WEDNESDAY

○ _____
○ _____
○ _____
○ _____
○ _____
○ _____
○ _____
○ _____

12 THURSDAY

○ _____
○ _____
○ _____
○ _____
○ _____

February
2026

13 FRIDAY

○ _____
○ _____
○ _____
○ _____
○ _____
○ _____
○ _____
○ _____

14 SATURDAY

○ _____
○ _____
○ _____
○ _____
○ _____
○ _____
○ _____
○ _____

15 SUNDAY

○ _____
○ _____
○ _____
○ _____
○ _____
○ _____
○ _____
○ _____

16 MONDAY

○ _____
○ _____
○ _____
○ _____
○ _____

17 TUESDAY

○
○
○
○
○
○
○
○

18 WEDNESDAY

○
○
○
○
○
○
○
○

19 THURSDAY

○
○
○
○
○
○
○
○

20 FRIDAY

○
○
○
○
○

February
2026

21 SATURDAY

22 SUNDAY

23 MONDAY

24 TUESDAY

25 WEDNESDAY

26 THURSDAY

27 FRIDAY

28 SATURDAY

March
2026

01 SUNDAY

○
○
○
○
○
○
○
○

02 MONDAY

○
○
○
○
○
○
○
○

03 TUESDAY

○
○
○
○
○
○
○
○

04 WEDNESDAY

○
○
○
○
○

05 THURSDAY

06 FRIDAY

07 SATURDAY

08 SUNDAY

March
2026

09 MONDAY

○ _____
○ _____
○ _____
○ _____
○ _____
○ _____
○ _____
○ _____

10 TUESDAY

○ _____
○ _____
○ _____
○ _____
○ _____
○ _____
○ _____
○ _____

11 WEDNESDAY

○ _____
○ _____
○ _____
○ _____
○ _____
○ _____
○ _____
○ _____

12 THURSDAY

○ _____
○ _____
○ _____
○ _____
○ _____

13 FRIDAY

14 SATURDAY

15 SUNDAY

16 MONDAY

17 TUESDAY

○ _____
○ _____
○ _____
○ _____
○ _____
○ _____
○ _____
○ _____

18 WEDNESDAY

○ _____
○ _____
○ _____
○ _____
○ _____
○ _____
○ _____
○ _____

19 THURSDAY

○ _____
○ _____
○ _____
○ _____
○ _____
○ _____
○ _____
○ _____

20 FRIDAY

○ _____
○ _____
○ _____
○ _____
○ _____

21 SATURDAY

22 SUNDAY

23 MONDAY

24 TUESDAY

25 WEDNESDAY

○ _____
○ _____
○ _____
○ _____
○ _____
○ _____
○ _____
○ _____

26 THURSDAY

○ _____
○ _____
○ _____
○ _____
○ _____
○ _____
○ _____
○ _____

27 FRIDAY

○ _____
○ _____
○ _____
○ _____
○ _____
○ _____
○ _____
○ _____

28 SATURDAY

○ _____
○ _____
○ _____
○ _____
○ _____

March
2026

29 SUNDAY
- ○
- ○
- ○
- ○
- ○
- ○
- ○
- ○

30 MONDAY
- ○
- ○
- ○
- ○
- ○
- ○
- ○
- ○

31 TUESDAY
- ○
- ○
- ○
- ○
- ○
- ○
- ○
- ○

NOTES

01 WEDNESDAY

○ _____
○ _____
○ _____
○ _____
○ _____
○ _____
○ _____
○ _____

02 THURSDAY

○ _____
○ _____
○ _____
○ _____
○ _____
○ _____
○ _____
○ _____

03 FRIDAY

○ _____
○ _____
○ _____
○ _____
○ _____
○ _____
○ _____
○ _____

04 SATURDAY

○ _____
○ _____
○ _____
○ _____
○ _____

April
2026

05 SUNDAY

○ _____
○ _____
○ _____
○ _____
○ _____
○ _____
○ _____
○ _____

06 MONDAY

○ _____
○ _____
○ _____
○ _____
○ _____
○ _____
○ _____
○ _____

07 TUESDAY

○ _____
○ _____
○ _____
○ _____
○ _____
○ _____
○ _____
○ _____

08 WEDNESDAY

○ _____
○ _____
○ _____
○ _____
○ _____

April
2026

09 THURSDAY

10 FRIDAY

11 SATURDAY

12 SUNDAY

13 MONDAY

14 TUESDAY

15 WEDNESDAY

16 THURSDAY

April
2026

17 FRIDAY

18 SATURDAY

19 SUNDAY

20 MONDAY

21 TUESDAY

22 WEDNESDAY

23 THURSDAY

24 FRIDAY

April
2026

25 SATURDAY

○
○
○
○
○
○
○

26 SUNDAY

○
○
○
○
○
○
○
○

27 MONDAY

○
○
○
○
○
○
○

28 TUESDAY

○
○
○
○
○

29 WEDNESDAY

○
○
○
○
○
○
○
○

30 THURSDAY

○
○
○
○
○
○
○
○

NOTES

May
2026

01 FRIDAY

○ _____
○ _____
○ _____
○ _____
○ _____
○ _____
○ _____
○ _____

02 SATURDAY

○ _____
○ _____
○ _____
○ _____
○ _____
○ _____
○ _____
○ _____

03 SUNDAY

○ _____
○ _____
○ _____
○ _____
○ _____
○ _____
○ _____
○ _____

04 MONDAY

○ _____
○ _____
○ _____
○ _____
○ _____

05 TUESDAY
○
○
○
○
○
○
○
○

06 WEDNESDAY
○
○
○
○
○
○
○
○

07 THURSDAY
○
○
○
○
○
○
○
○

08 FRIDAY
○
○
○
○
○

May
2026

09 SATURDAY

10 SUNDAY

11 MONDAY

12 TUESDAY

May
2026

13 WEDNESDAY

14 THURSDAY

15 FRIDAY

16 SATURDAY

17 SUNDAY

○ _____
○ _____
○ _____
○ _____
○ _____
○ _____
○ _____
○ _____

18 MONDAY

○ _____
○ _____
○ _____
○ _____
○ _____
○ _____
○ _____
○ _____

19 TUESDAY

○ _____
○ _____
○ _____
○ _____
○ _____
○ _____
○ _____
○ _____

20 WEDNESDAY

○ _____
○ _____
○ _____
○ _____
○ _____

May
2026

21 THURSDAY

22 FRIDAY

23 SATURDAY

24 SUNDAY

May
2026

25 MONDAY

- ○
- ○
- ○
- ○
- ○
- ○
- ○
- ○

26 TUESDAY

- ○
- ○
- ○
- ○
- ○
- ○
- ○
- ○

27 WEDNESDAY

- ○
- ○
- ○
- ○
- ○
- ○
- ○
- ○

28 THURSDAY

- ○
- ○
- ○
- ○
- ○

May
2026

29 FRIDAY

○ _____
○ _____
○ _____
○ _____
○ _____
○ _____
○ _____
○ _____

30 SATURDAY

○ _____
○ _____
○ _____
○ _____
○ _____
○ _____
○ _____
○ _____

31 SUNDAY

○ _____
○ _____
○ _____
○ _____
○ _____
○ _____
○ _____
○ _____

NOTES

_____
_____
_____
_____

June 2026

01 MONDAY
- ○
- ○
- ○
- ○
- ○
- ○
- ○
- ○

02 TUESDAY
- ○
- ○
- ○
- ○
- ○
- ○
- ○
- ○

03 WEDNESDAY
- ○
- ○
- ○
- ○
- ○
- ○
- ○
- ○

04 THURSDAY
- ○
- ○
- ○
- ○
- ○

05 FRIDAY

06 SATURDAY

07 SUNDAY

08 MONDAY

09 TUESDAY

○ _____
○ _____
○ _____
○ _____
○ _____
○ _____
○ _____
○ _____

10 WEDNESDAY

○ _____
○ _____
○ _____
○ _____
○ _____
○ _____
○ _____
○ _____
○ _____

11 THURSDAY

○ _____
○ _____
○ _____
○ _____
○ _____
○ _____
○ _____
○ _____

12 FRIDAY

○ _____
○ _____
○ _____
○ _____
○ _____

13 SATURDAY

14 SUNDAY

15 MONDAY

16 TUESDAY

17 WEDNESDAY

- ○
- ○
- ○
- ○
- ○
- ○
- ○
- ○

18 THURSDAY

- ○
- ○
- ○
- ○
- ○
- ○
- ○
- ○

19 FRIDAY

- ○
- ○
- ○
- ○
- ○
- ○
- ○
- ○

20 SATURDAY

- ○
- ○
- ○
- ○
- ○

21 SUNDAY

22 MONDAY

23 TUESDAY

24 WEDNESDAY

25 THURSDAY

26 FRIDAY

27 SATURDAY

28 SUNDAY

29 MONDAY

○ _____
○ _____
○ _____
○ _____
○ _____
○ _____
○ _____
○ _____

30 TUESDAY

○ _____
○ _____
○ _____
○ _____
○ _____
○ _____
○ _____
○ _____

NOTES

July
2026

01 WEDNESDAY

02 THURSDAY

03 FRIDAY

04 SATURDAY

July
2026

05 SUNDAY

○
○
○
○
○
○
○

06 MONDAY

○
○
○
○
○
○
○
○

07 TUESDAY

○
○
○
○
○
○
○
○

08 WEDNESDAY

○
○
○
○
○

09 THURSDAY

○ _____
○ _____
○ _____
○ _____
○ _____
○ _____
○ _____
○ _____

10 FRIDAY

○ _____
○ _____
○ _____
○ _____
○ _____
○ _____
○ _____
○ _____

11 SATURDAY

○ _____
○ _____
○ _____
○ _____
○ _____
○ _____
○ _____
○ _____

12 SUNDAY

○ _____
○ _____
○ _____
○ _____
○ _____

13 MONDAY

○
○
○
○
○
○
○
○

14 TUESDAY

○
○
○
○
○
○
○
○

15 WEDNESDAY

○
○
○
○
○
○
○
○

16 THURSDAY

○
○
○
○
○

17 FRIDAY

○ _____
○ _____
○ _____
○ _____
○ _____
○ _____
○ _____
○ _____

18 SATURDAY

○ _____
○ _____
○ _____
○ _____
○ _____
○ _____
○ _____
○ _____

19 SUNDAY

○ _____
○ _____
○ _____
○ _____
○ _____
○ _____
○ _____
○ _____

20 MONDAY

○ _____
○ _____
○ _____
○ _____
○ _____

21 TUESDAY

○ _____
○ _____
○ _____
○ _____
○ _____
○ _____
○ _____
○ _____

22 WEDNESDAY

○ _____
○ _____
○ _____
○ _____
○ _____
○ _____
○ _____
○ _____

23 THURSDAY

○ _____
○ _____
○ _____
○ _____
○ _____
○ _____
○ _____
○ _____

24 FRIDAY

○ _____
○ _____
○ _____
○ _____
○ _____

25 SATURDAY

○ _____
○ _____
○ _____
○ _____
○ _____
○ _____
○ _____
○ _____

26 SUNDAY

○ _____
○ _____
○ _____
○ _____
○ _____
○ _____
○ _____
○ _____

27 MONDAY

○ _____
○ _____
○ _____
○ _____
○ _____
○ _____
○ _____
○ _____

28 TUESDAY

○ _____
○ _____
○ _____
○ _____
○ _____

July
2026

29 WEDNESDAY

○
○
○
○
○
○
○
○

30 THURSDAY

○
○
○
○
○
○
○
○

31 FRIDAY

○
○
○
○
○
○
○
○

NOTES

August
2026

01 SATURDAY

○
○
○
○
○
○
○
○

02 SUNDAY

○
○
○
○
○
○
○
○

03 MONDAY

○
○
○
○
○
○
○
○

04 TUESDAY

○
○
○
○
○

05 WEDNESDAY

06 THURSDAY

07 FRIDAY

08 SATURDAY

August
2026

09 SUNDAY

10 MONDAY

11 TUESDAY

12 WEDNESDAY

13 THURSDAY

14 FRIDAY

15 SATURDAY

16 SUNDAY

August
2026

17 MONDAY

18 TUESDAY

19 WEDNESDAY

20 THURSDAY

21 FRIDAY

○
○
○
○
○
○
○
○

22 SATURDAY

○
○
○
○
○
○
○
○

23 SUNDAY

○
○
○
○
○
○
○
○

24 MONDAY

○
○
○
○
○

25 TUESDAY

○
○
○
○
○
○
○
○

26 WEDNESDAY

○
○
○
○
○
○
○
○

27 THURSDAY

○
○
○
○
○
○
○
○

28 FRIDAY

○
○
○
○
○

29 SATURDAY

30 SUNDAY

31 MONDAY

NOTES

September
2026

01 TUESDAY
- ○
- ○
- ○
- ○
- ○
- ○
- ○
- ○

02 WEDNESDAY
- ○
- ○
- ○
- ○
- ○
- ○
- ○
- ○

03 THURSDAY
- ○
- ○
- ○
- ○
- ○
- ○
- ○
- ○

04 FRIDAY
- ○
- ○
- ○
- ○
- ○

05 SATURDAY

○ _____
○ _____
○ _____
○ _____
○ _____
○ _____
○ _____
○ _____

06 SUNDAY

○ _____
○ _____
○ _____
○ _____
○ _____
○ _____
○ _____
○ _____

07 MONDAY

○ _____
○ _____
○ _____
○ _____
○ _____
○ _____
○ _____
○ _____

08 TUESDAY

○ _____
○ _____
○ _____
○ _____
○ _____

09 WEDNESDAY

10 THURSDAY

11 FRIDAY

12 SATURDAY

13 SUNDAY

○
○
○
○
○
○
○
○

14 MONDAY

○
○
○
○
○
○
○
○

15 TUESDAY

○
○
○
○
○
○
○
○

16 WEDNESDAY

○
○
○
○
○

17 THURSDAY

○
○
○
○
○
○
○
○

18 FRIDAY

○
○
○
○
○
○
○
○

19 SATURDAY

○
○
○
○
○
○
○
○

20 SUNDAY

○
○
○
○
○

21 MONDAY
- ○
- ○
- ○
- ○
- ○
- ○
- ○
- ○

22 TUESDAY
- ○
- ○
- ○
- ○
- ○
- ○
- ○
- ○

23 WEDNESDAY
- ○
- ○
- ○
- ○
- ○
- ○
- ○
- ○

24 THURSDAY
- ○
- ○
- ○
- ○
- ○

25 FRIDAY

○ _____
○ _____
○ _____
○ _____
○ _____
○ _____
○ _____
○ _____

26 SATURDAY

○ _____
○ _____
○ _____
○ _____
○ _____
○ _____
○ _____
○ _____

27 SUNDAY

○ _____
○ _____
○ _____
○ _____
○ _____
○ _____
○ _____

28 MONDAY

○ _____
○ _____
○ _____
○ _____
○ _____

29 TUESDAY

- ○
- ○
- ○
- ○
- ○
- ○
- ○
- ○

30 WEDNESDAY

- ○
- ○
- ○
- ○
- ○
- ○
- ○
- ○

NOTES

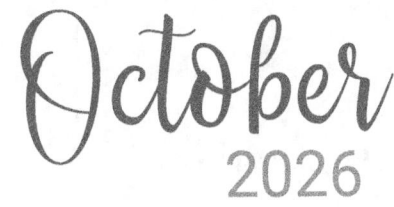

October
2026

01 THURSDAY

02 FRIDAY

03 SATURDAY

04 SUNDAY

October
2026

05 MONDAY

06 TUESDAY

07 WEDNESDAY

08 THURSDAY

09 FRIDAY

○ _____
○ _____
○ _____
○ _____
○ _____
○ _____
○ _____
○ _____

10 SATURDAY

○ _____
○ _____
○ _____
○ _____
○ _____
○ _____
○ _____
○ _____

11 SUNDAY

○ _____
○ _____
○ _____
○ _____
○ _____
○ _____
○ _____
○ _____

12 MONDAY

○ _____
○ _____
○ _____
○ _____
○ _____

13 TUESDAY

14 WEDNESDAY

15 THURSDAY

16 FRIDAY

17 SATURDAY

18 SUNDAY

19 MONDAY

20 TUESDAY

21 WEDNESDAY

22 THURSDAY

23 FRIDAY

24 SATURDAY

October
2026

25 SUNDAY
○
○
○
○
○
○
○
○

26 MONDAY
○
○
○
○
○
○
○
○

27 TUESDAY
○
○
○
○
○
○
○
○

28 WEDNESDAY
○
○
○
○
○

29 THURSDAY

30 FRIDAY

31 SATURDAY

NOTES

November
2026

01 SUNDAY

02 MONDAY

03 TUESDAY

04 WEDNESDAY

05 THURSDAY

06 FRIDAY

07 SATURDAY

08 SUNDAY

November
2026

09 MONDAY

10 TUESDAY

11 WEDNESDAY

12 THURSDAY

13 FRIDAY

14 SATURDAY

15 SUNDAY

16 MONDAY

November
2026

17 TUESDAY

18 WEDNESDAY

19 THURSDAY

20 FRIDAY

21 SATURDAY

22 SUNDAY

23 MONDAY

24 TUESDAY

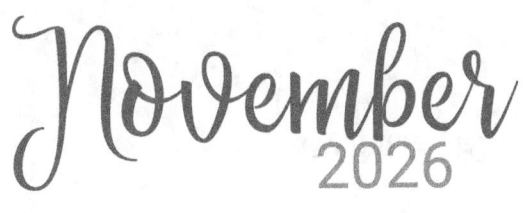

25 WEDNESDAY

26 THURSDAY

27 FRIDAY

28 SATURDAY

29 SUNDAY

○ _____
○ _____
○ _____
○ _____
○ _____
○ _____
○ _____
○ _____

30 MONDAY

○ _____
○ _____
○ _____
○ _____
○ _____
○ _____
○ _____
○ _____

NOTES

_____
_____
_____
_____
_____
_____
_____
_____
_____
_____
_____
_____

01 TUESDAY

○ _____
○ _____
○ _____
○ _____
○ _____
○ _____
○ _____
○ _____

02 WEDNESDAY

○ _____
○ _____
○ _____
○ _____
○ _____
○ _____
○ _____
○ _____

03 THURSDAY

○ _____
○ _____
○ _____
○ _____
○ _____
○ _____
○ _____
○ _____

04 FRIDAY

○ _____
○ _____
○ _____
○ _____
○ _____

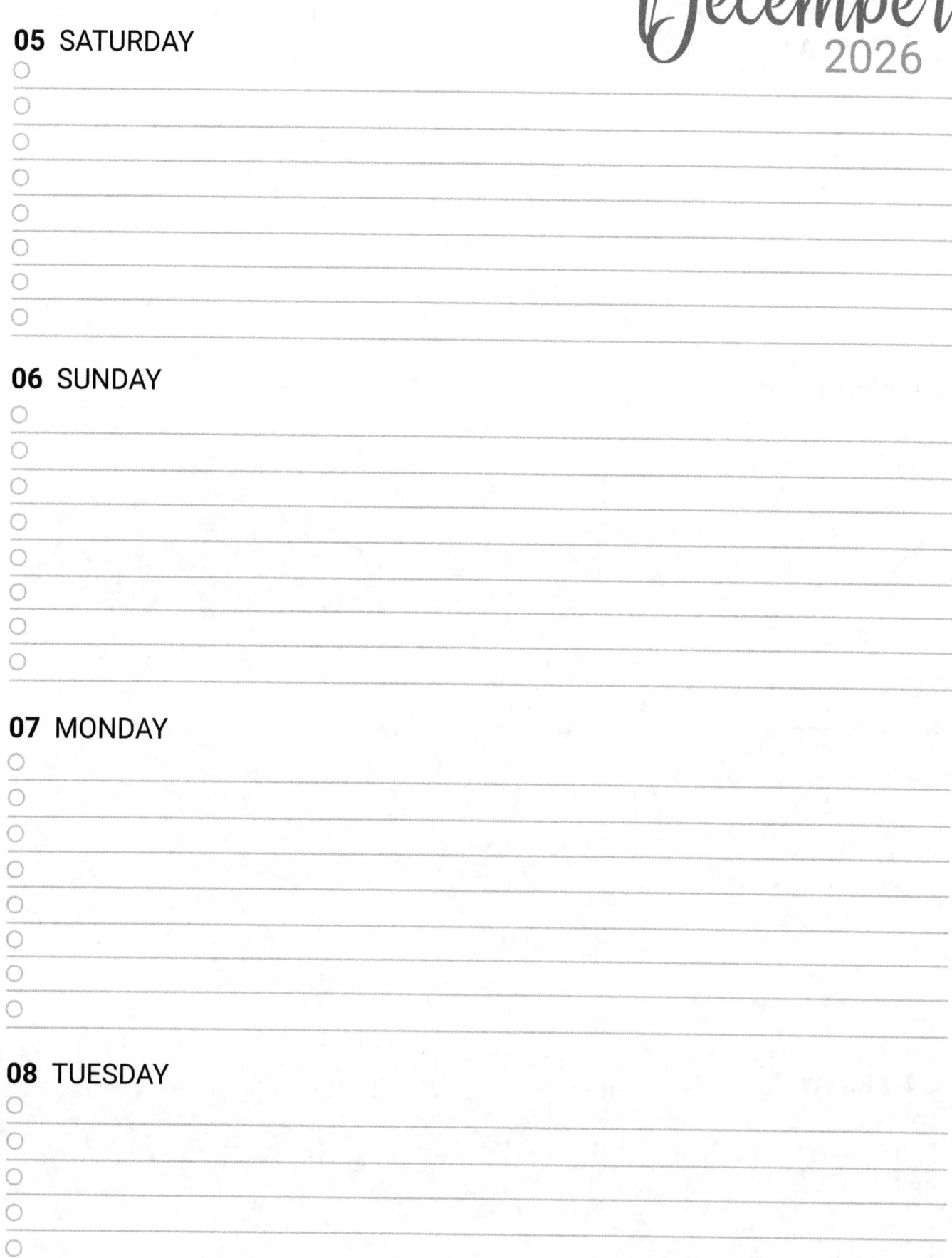

December
2026

05 SATURDAY

06 SUNDAY

07 MONDAY

08 TUESDAY

09 WEDNESDAY

○

○

○

○

○

○

○

10 THURSDAY

○

○

○

○

○

○

○

○

11 FRIDAY

○

○

○

○

○

○

○

○

12 SATURDAY

○

○

○

○

○

13 SUNDAY

14 MONDAY

15 TUESDAY

16 WEDNESDAY

17 THURSDAY

○
○
○
○
○
○
○
○

18 FRIDAY

○
○
○
○
○
○
○
○

19 SATURDAY

○
○
○
○
○
○
○
○

20 SUNDAY

○
○
○
○
○

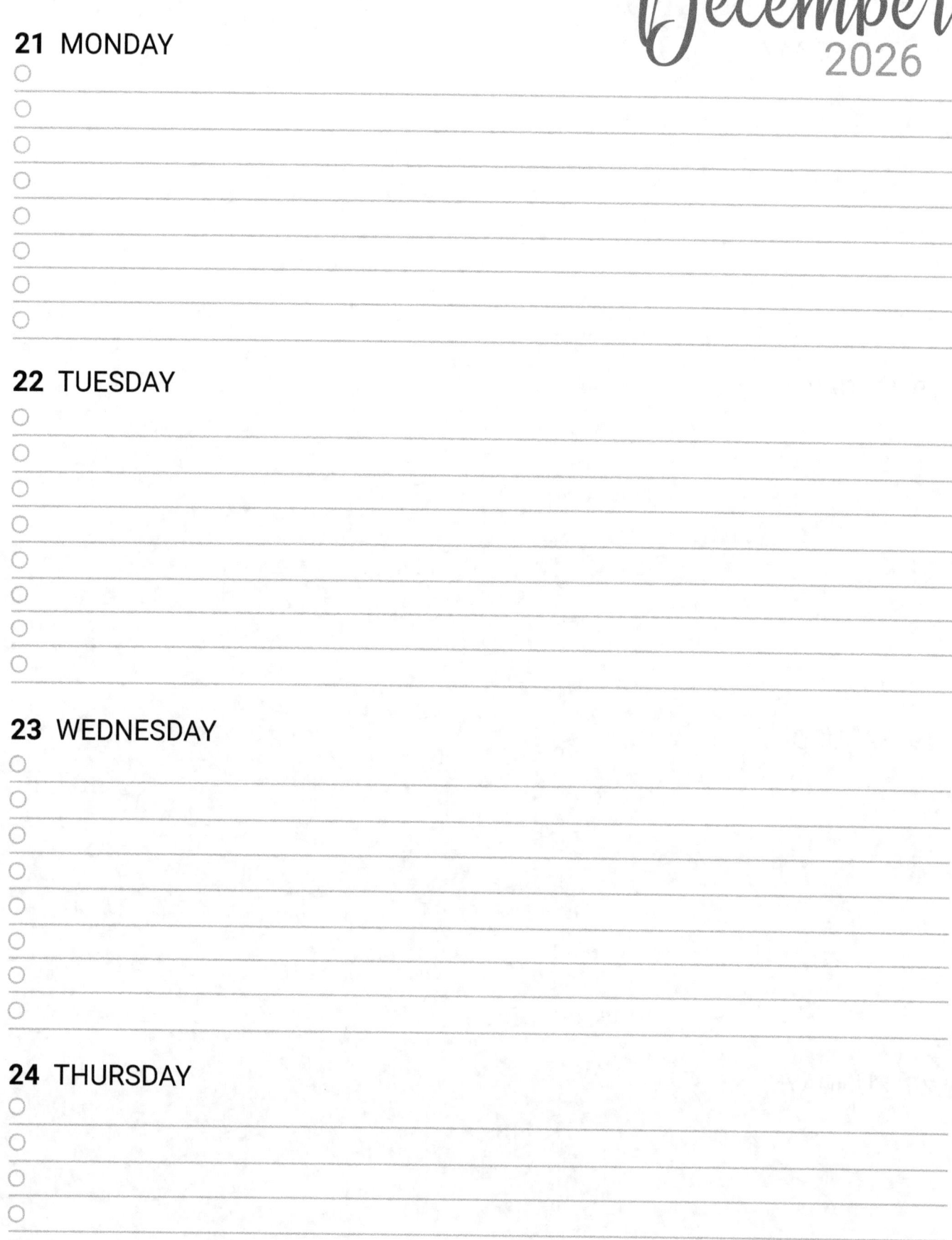

December
2026

21 MONDAY

○
○
○
○
○
○
○
○

22 TUESDAY

○
○
○
○
○
○
○
○

23 WEDNESDAY

○
○
○
○
○
○
○
○

24 THURSDAY

○
○
○
○
○

25 FRIDAY

○
○
○
○
○
○
○
○

26 SATURDAY

○
○
○
○
○
○
○
○

27 SUNDAY

○
○
○
○
○
○
○
○

28 MONDAY

○
○
○
○
○

December
2026

29 TUESDAY

○
○
○
○
○
○
○

30 WEDNESDAY

○
○
○
○
○
○
○
○

31 THURSDAY

○
○
○
○
○
○
○

NOTES